MATH PUZZLES

MATH PUZZLES

BY PEGGY ADLER
AND IRVING ADLER

Illustrated by Peggy Adler

Franklin Watts
New York □ London □ 1978

Library of Congress Cataloging in Publication Data

Adler, Peggy.
Math puzzles.

SUMMARY: A collection of math puzzles involving simple algebra, prime numbers, and other aspects of arithmetic interspersed with line drawings.

1. Mathematical recreations—Juvenile literature. [1. Mathematical recreations] I. Adler, Irving, joint author. II. Title.

QA95.A25 793.7'4 78–2833

ISBN 0–531–02216–1

Printed in the United States of America
6 5 4

CONTENTS

INTRODUCTION

Have you always been a little bit afraid of math problems or puzzles? Has a problem ever seemed so difficult that you didn't want to spend any time trying to solve it? Many people feel this way. But they don't have to. Solving math puzzles can be fun. And this book has enough different types of math puzzles to give you many delightful moments.

If you get stumped on any of the puzzles, you can always turn to the back of the book, and learn, step-by-step, how to arrive at the answer. You may even find that you have learned so many new ways of solving math problems, that your skill as well as your enjoyment of math has increased.

PUZZLES

1 □ Wendy has just skied down to the bottom of a hill in her back yard. Her skis made a trail 360 feet long, from the top of the hill to the bottom.

Now she is going to climb back to the top of the hill, along the trail that her skis made when she skied down.

Every minute that Wendy climbs up the hill, she moves forward 3 yards during the first 30 seconds, and slides back 2 yards while she rests during the next 30 seconds.

How long will it take her to reach the top of the hill?

2 □ One day Davey went mountain climbing. He hiked along the Blue Dot Trail up Mt. Minuit at the rate of 3 miles per hour. No sooner had he reached the top of Mt. Minuit than he immediately began his return trip down the mountain. He went down Mt. Minuit along the Blue Dot Trail at a rate of 6 miles per hour. What was the average speed of Davey's round trip?

3 □ Naomi took a bus from her home in Montgomery to her grandparents' house in Shaftsbury. The bus traveled at an average speed of 30 miles per hour. How fast would the bus have to travel on the return trip to Montgomery in order for the average speed of the round trip to be 60 miles per hour?

4 □ One beautiful day in May, Amy decided to hike from Branford to Madison along the rugged Hammonasset Trail. The only town that she would pass through along the way is Guilford.

Forty minutes after Amy left Branford she came upon a sign nailed to a tree. Amy stopped, and read the sign. It said, "If you have come from Branford, then you have already gone half as far as it is from here to Guilford." Amy rested beneath the tree and ate a pastrami sandwich that she had brought with her. Then, feeling refreshed, she continued on her hike.

Amy trekked on for another 11 miles, at the end of which she came upon a second sign nailed to a tree. Once again she stopped to read the sign. This sign said, "If you are going to Madison, then you still have to go half as far as it is from here to Guilford." Here, too, Amy rested beneath the tree, this time eating a yellow delicious apple and some homemade fudge. Then, once again, feeling refreshed, Amy continued on her way. She hiked on, along the Hammonasset Trail for one more hour, and at long last reached Madison.

If Amy hiked at her own steady pace all the way, what is the length of the Hammonasset Trail between Branford and Madison?

5 □ When Wendy visited New Orleans, Louisiana, she went on a bayou tour on a riverboat. The riverboat left the Canal Street dock in New Orleans at 11:00 A.M. It cruised 10 miles down the Mississippi River, until it came to an intracoastal waterway known as the Algiers Route. The riverboat turned into the Algiers Route. The level of the water in the Algiers Route is lower than the level of the water in the Mississippi River. Therefore, the riverboat had to enter the Algiers Lock, to be lowered to the waterway's water level. The riverboat entered the lock, and the gate closed behind it. The lock gate in front of the riverboat was already closed. In 10 minutes, the level of the water in the lock had been lowered to the level of that in the Algiers Route. The gate at the front of the lock opened, and the riverboat glided out into the waterway.

The riverboat cruised 7 miles along the Algiers Route until it came to Hero Cutoff. At Hero Cutoff, it turned into a canal known as the Bayou Barataria. There Wendy saw fishermen painting their boats and drying their nets. On either side of the canal there were moss-covered oaks and magnolia trees, as well as cypress, palmettos, and willows.

The riverboat glided 14 miles along the Bayou Barataria, through the bayous where the Pirate Jean Lafitte and his buccaneers had once lived, until it came to an old Indian burial ground. One side of the burial ground was bordered by the canal. The rest of it was surrounded by a dense forest. The Indians' graves all lay at the foot of a gigantic earth mound. An oak tree that was hundreds of years old grew on top of the mound. At this point in the canal, the riverboat turned around and immediately began its return trip to New Orleans.

The riverboat cruised back along the Bayou Barataria, to Hero Cutoff. Here, it turned into the Harvey Canal, a 3-mile waterway that leads back to the Mississippi River. The level

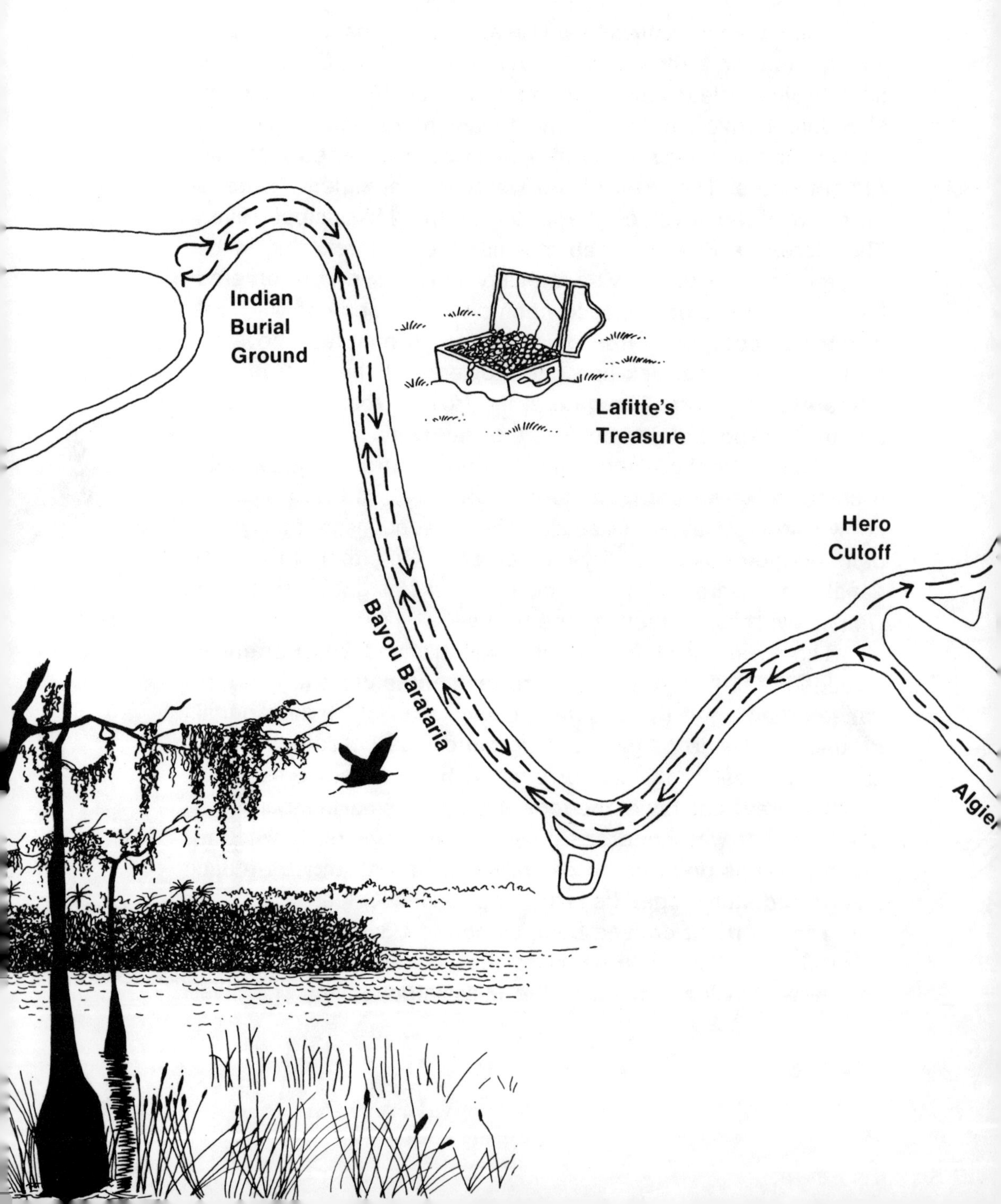
Indian
Burial
Ground
Lafitte's
Treasure
Bayou Barataria
Hero
Cutoff

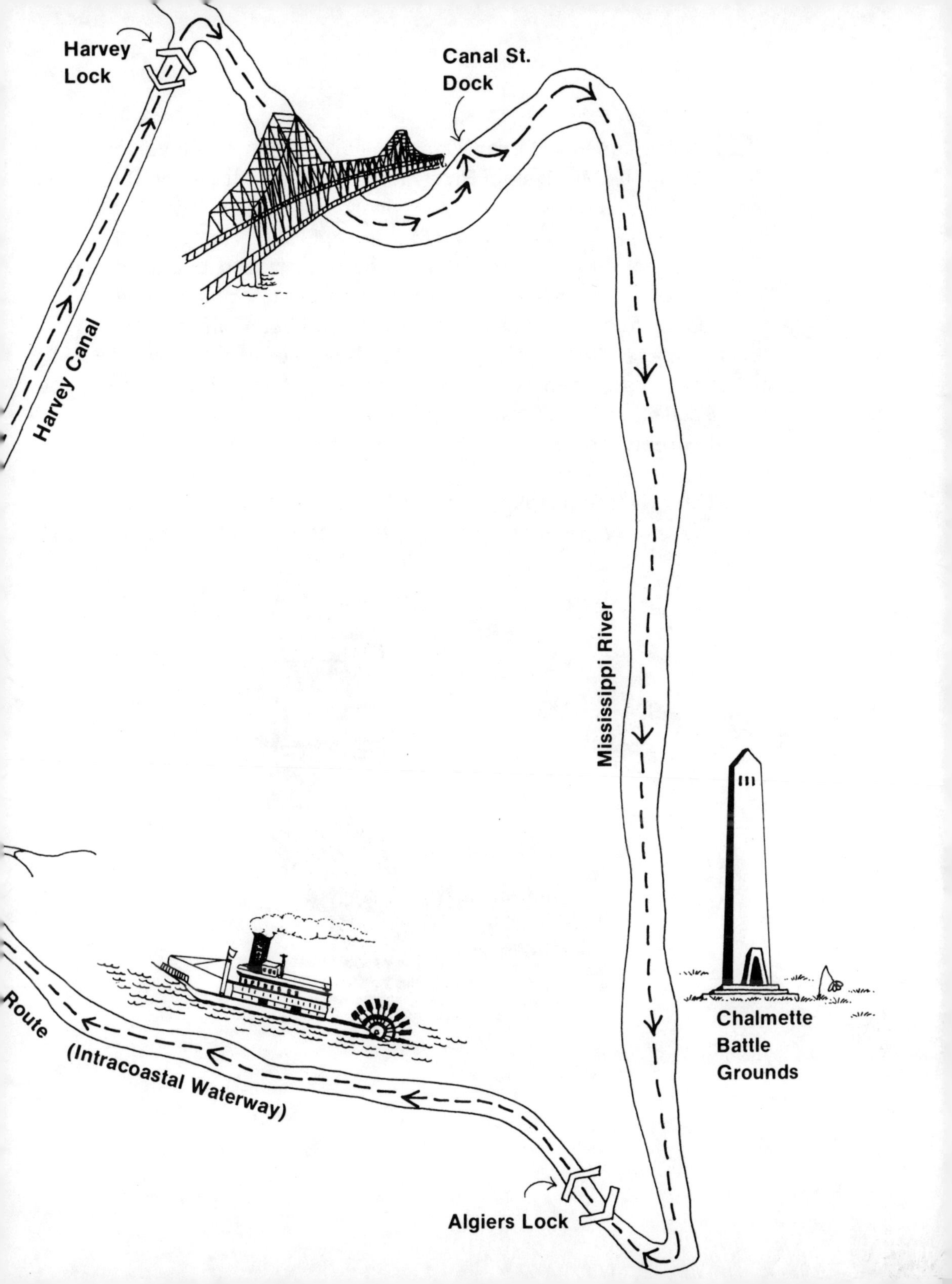
Harvey
Lock
Canal St.
Dock
Harvey Canal
Mississippi River
Chalmette
Battle
Grounds
Algiers Lock
Route
(Intracoastal Waterway)

of the water in the Harvey Canal is lower than the level of the water in the Mississippi River. Therefore, after traveling the length of the Harvey Canal, the riverboat had to enter the Harvey Lock, to be raised to the Mississippi River's water level. The riverboat entered the lock and the gate closed behind it. The lock gate in front of the riverboat was already closed. In 10 minutes the level of the water in the lock had been raised to the level of that in the Mississippi River. The gate at the front of the lock opened, and the riverboat glided out and turned into the Mississippi. After traveling 4 more miles, it reached the Canal Street dock from which it had departed earlier in the day.

If the riverboat traveled at an average speed of 10 miles per hour, at what time did Wendy arrive back in New Orleans?

6 □ One sunny day, Marty went for a walk in the woods. He hadn't gone far when, through his binoculars, he saw Bali in the distance up in a tree. She was hammering nails into a tree-house that she was making. Marty noticed that Bali hit a nail with the hammer 30 times every minute. He also realized that every time that he saw Bali swing the hammer to hit a nail, he heard a banging sound. After Bali stopped hammering Marty heard one more bang.

If light travels at a speed of 186,000 miles per second, and sound travels at a speed of 1,100 feet per second, how far was Marty from Bali?

7 □ This afternoon Amy ran in a one-mile race. She ran the first three-quarters of a mile in 6-3/4 minutes. Amy also ran the first half-mile in exactly the same amount of time that it took her to run the second half-mile. Furthermore, she ran the third quarter of the race in the same time as she ran the last quarter. How long did it take for Amy to run the whole mile?

8 □ Charlotte and Amy are picking strawberries. If Charlotte picks 6 dozen dozen strawberries, and Amy picks a half-dozen dozen strawberries, which hedgehog will have picked the larger crop of strawberries?

9 □ Early one morning, just as the sun was rising in the eastern sky, Davey went out into the fields near his house to pick mushrooms. He hadn't been picking long when his friend Amy came by on her daily morning walk. Davey, being a very generous fellow, told Amy that that night he would give her 2/3 of his day's pick, keeping 1/3 for himself. Amy thanked him and went happily on her way. Later that same day, along came Davey's other friend, Meredith. Davey, being as absentminded as he is generous, told Meredith that that night he would give her 1/3 of his day's pick, keeping 2/3 for himself.

That evening, as the sun was setting in the western sky, Davey returned home. When he arrived at his house, Amy and

Meredith were waiting for him. "Good grief!" said Davey to himself. "What have I done! How can I ever keep my promises to my friends, and still have some mushrooms left for myself?"

Can you find a way in which Davey can fairly divide the mushrooms, so that each of the three hedgehogs still receives his or her intended portion, in best accordance with Davey's promises?

10 □ Find four different fractions, in each of which the numerator is one less than the denominator, so that when all four of the fractions are added together, their sum will be a whole number.

11 □ Bali is sawing a strip of wood that is 50 inches long into pieces that are 1 inch long. If it takes one minute to saw off each 1-inch piece, then how many minutes will it take to cut up the entire 50-inch strip?

12 □ Two fathers and 2 sons divided 3 apples among themselves. Each one received exactly 1 apple. How is this possible?

13 □ Charlotte and her friends are having a mushroom party. The 5 hedgehogs are able to eat 5 mushrooms in 5 minutes. How many hedgehogs will it take to eat 100 mushrooms in 100 minutes, if they all eat at the same rate?

14 □ Jonathan has 21 jugs. Some of his jugs have molasses in them. Seven of his jugs are full of molasses. Three of his jugs are half-full. Four of his jugs are half-empty. And 7 of his jugs contain no molasses at all.

Jonathan is a hedgehog with a very big heart. He believes in sharing. He is going to share his jugs with his brother Nathaniel and his friend Meredith.

Jonathan is going to divide his jugs among the three of them in the following manner: Each of the three hedgehogs will receive the same number of jugs; each hedgehog will also receive an equal amount of molasses; no molasses will be transferred from jug to jug; and none of the three hedgehogs will have as many as 4 jugs of the same description (full, half-full, half-empty, or empty).

How does Jonathan succeed in making a correct division of his jugs?

15 □ Jonathan took one full glass of orange juice from a flask that contained one pint of orange juice. He poured the glassful of orange juice into a jug that contained one pint of apple cider.

Next Jonathan took one glassful of the mixture of orange juice and apple cider from the jug. He poured this glassful of the mixture of the two beverages into the flask containing the orange juice.

In the end, had Jonathan transferred more orange juice to the jug, or more cider to the flask?

16 □ Naomi has two dogs, named Paxi and Blixi, that are equal in weight.

Paxi has just been swimming in the brook that runs through Naomi's back yard. Now Paxi's weight is equal to that of Blixi and one 3/4-pound weight. If Blixi's dry weight is three-fourths of Paxi's wet weight, then how much does each dog weigh when dry?

17 □ One day, Marty said to Bali, "Select any three-digit number (abc) and repeat it (abcabc), and I will guarantee you four things that will be true about your six-digit number."

1. "Your six digit number will be divisible by 13."
2. "The quotient that you arrived at when you divided your six-digit number by 13 will itself be divisible by 11."
3. "The quotient that you arrived at after you divided by 11 will itself be divisible by 7."
4. "The quotient that you will arrive at after you have divided by 7 will be your original three-digit number."

Why?

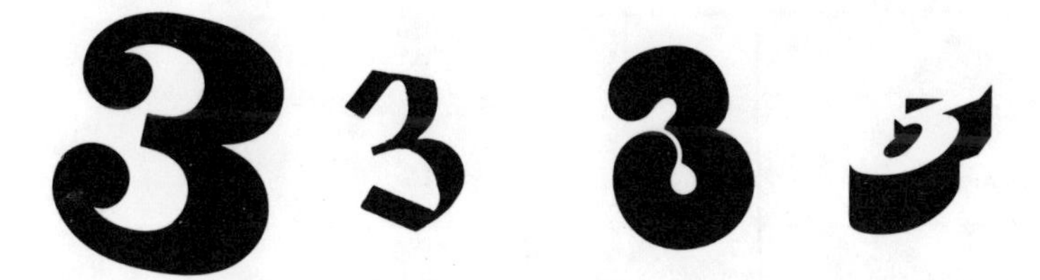

18 □ (5 + 5) × (5 + 5) = 100. Here are four fives, arranged in such a way that, using the arithmetical signs shown, the result of the computation is 100. Now arrange four threes, using any combination of the same arithmetical signs, or any others, so that the result of that computation, too, is 100.

19 □ Now arrange four nines, using any combination of arithmetical signs, as described in puzzle number 18, so that the result of the computation is 20.

20 □ Write an even number using only odd digits.

21 □ Write a hundred without using any zeros.

22 □ How many quarter-inch squares does it take to make an inch square?

23 □ Draw a square on a piece of paper. Divide the square into 9 boxes arranged in three horizontal rows and three vertical columns. Put one of the numbers from 1 to 9, inclusive, into each of the 9 boxes in the square. Use each number only once. Arrange the numbers so that the three numbers in each of the three horizontal rows, and in each of the three vertical columns, and in each of the two diagonals add up to the same sum. This type of arrangement is called a *magic square.*

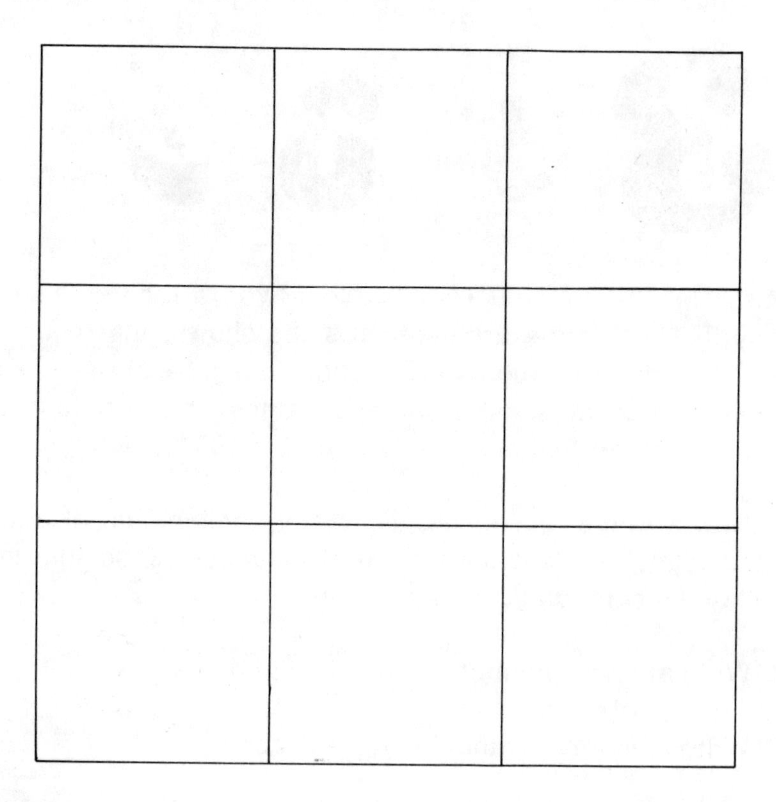

24 ☐ Join each prime number to the next higher prime number. (A prime number is a number that is only divisible by itself or by one.)

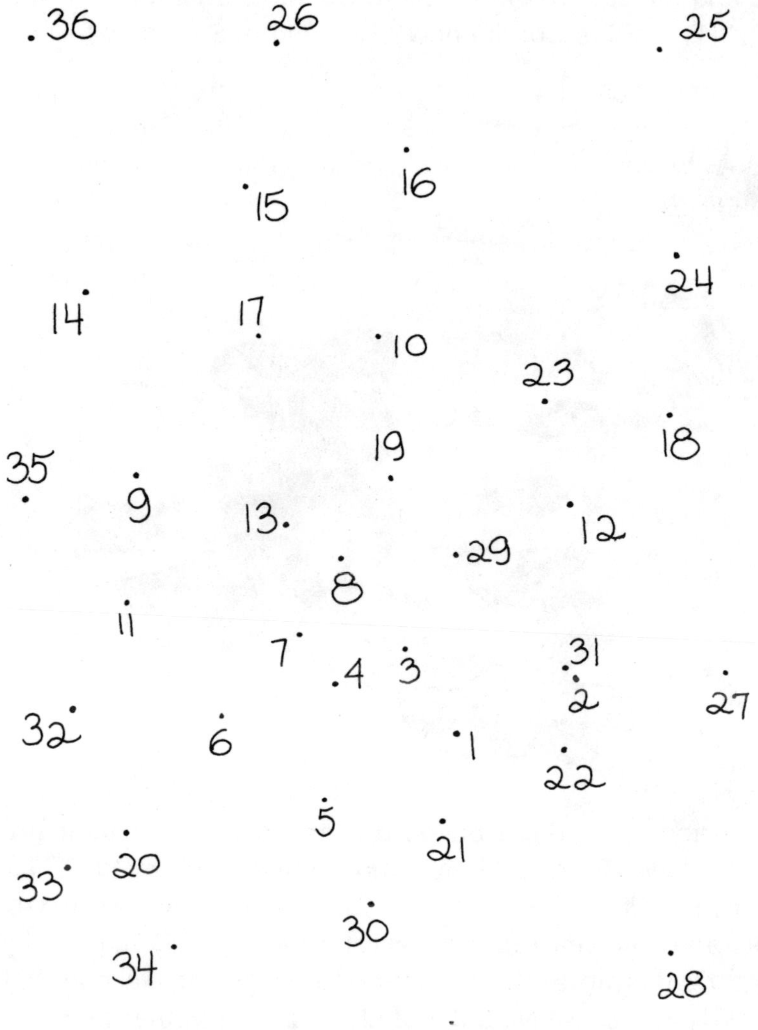

.36 26 .25

15 16

14 17 24

10 23

35 9 19 18

13. 29 12

8

11

7 .4 3 31

32 6 2 27

1

22

5

33 20 21

30

34

28

25 □ One day, not too long ago, Bali asked Connie and Sam their ages. Sam thought for a while, and then replied, "After our next birthdays have come and gone, the square of Connie's age, added to my age, will be 135; on the other hand, if you add the square of my age to Connie's age, the sum will be 207." How old is Connie now? How old is Sam now?

26 □ Naomi and Marty baked a birthday cake for their grandfather. Their friend Wendy, who helped them make it, asked the two children, "How old will your grandfather be this year?" "His age," Naomi replied, "will be between 50 and 79 inclusive. Furthermore, the number of divisors of his age will be odd." How old will Naomi and Marty's grandfather be?

27 □ Bali has set up a refreshment stand. She is selling lemonade and iced tea. Bali is charging the same price for each glass that she sells of either of the two drinks. At the end of the first

day, Bali found that she had sold just as much lemonade as she had sold iced tea. She also found that her lemonade sales had produced a profit of 10%, that is, 10¢ profit for every 100¢ of cost, or 1¢ of profit for every 10¢ of cost. Furthermore, she found that her iced tea sales had produced a loss of 10%, that is, 10¢ of loss for every 100¢ of cost, or 1¢ of loss for every 10¢ of cost. Did Bali's refreshment stand operate at a profit or a loss, or did she break even?

28 □ Sam went to the store to buy himself a new pair of jeans. When he got there, he found that they were having a clearance sale. He bought himself a pair of faded blue jeans that had

been reduced in price on two separate occasions. The price tag revealed that when the pants were first put on sale, the original purchase price had been crossed out, and the new price had been written in beneath it, revealing a discount of 10%. The tag further revealed that this first discount price had also been crossed out, and that the price of the pants had been further reduced by another 10%. What single discount is the same as these two successive discounts of 10%?

29 □ When Connie was born, her grandfather deposited $2,000 into a savings account that he was starting for her college education. Every three years after that, on her birthday, he deposited a sum of money that, when added to the interest that had accumulated during those three years, increased the amount of money in the account by half of it.

How much money was in Connie's college fund on her eighteenth birthday?

30 □ Connie has seven small silver chains. Each small chain has 5 links. She has gone to a silversmith to have the seven small chains linked together to form one long chain. The silversmith charges 25¢ for each link that he must cut open. He also charges 50¢ for each link that he solders closed.

What is the least amount of money that Connie will have to spend to have the seven small chains linked together to form one long chain?

31 □ Bali had a sum of money in her pockets. All of the money was in common United States coins. She did not have any silver dollars. She could not make change for a dollar, a half-dollar, a quarter, a dime, or a nickel.

What is the *largest* sum of money that Bali could have had in her pockets?

32 □ Charlotte and Amy have 110 pennies between them. If Charlotte has $1 more than Amy, then how many pennies does Charlotte have? How many pennies does Amy have?

33 □ One hot summer day, Bali and Marty went to the corner store to buy some ice cream.

Bali, who had a small appetite, decided to get an ice cream cone. Marty, who loved to eat sweets, decided to order a banana split.

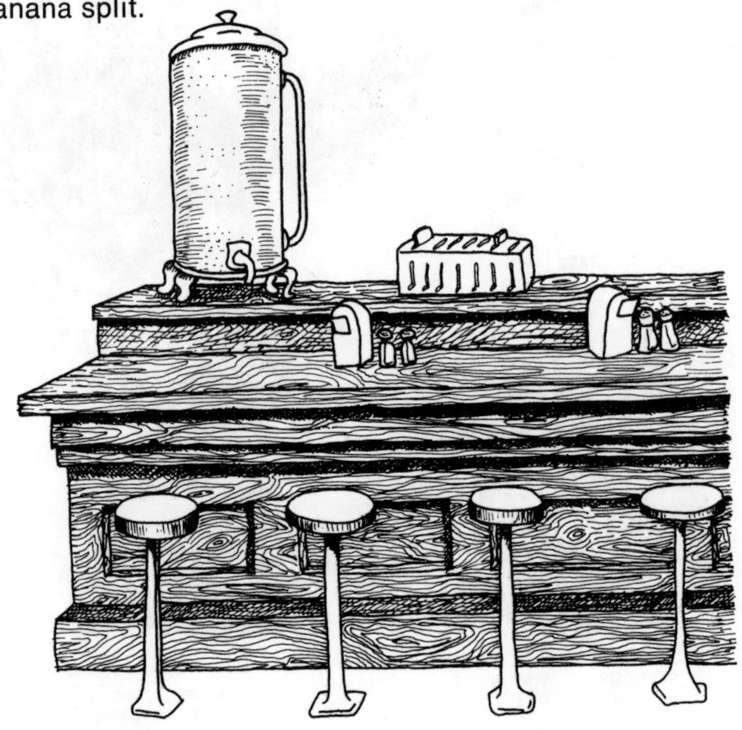

Since the two children had just one dollar between them, they asked the countergirl how much each of their selections would cost. "A banana split," she replied, "will buy you three ice cream cones. Furthermore," said the young woman, "the sum of the digits of the amount that you will owe me is 14."

The children found that they could, indeed, afford their selections, and still have some money left over.

How much did Marty's banana split cost?

34 □ Wendy is taking the 2:00 P.M. train from New York to New Haven. Every hour, on the hour, a train leaves New York to go to New Haven. Every hour, on the hour, another train leaves New Haven to go to New York. If the train ride between the two cities takes exactly two hours, how many trains will Wendy see that are going in the opposite direction?

35 □ Connie and Sam are planning to go to the movies this afternoon. They would like to know at what times the feature movie begins. Connie called the movie theater and was told, "The first showing of the feature movie begins at noon, when the hour hand and the minute hand are together. The next showing of the feature movie will begin the very next time that these two hands are together." At what time does the second afternoon showing of the feature movie begin?

36 □ The Island of Quonk, which lies somewhere between Here and There, is inhabited by two tribes, the Trolls and the Ffaffnirs. The Trolls are renowned for their honesty, since they *always tell the truth.* And the Ffaffnirs are equally well known for being habitual liars, since they *never tell the truth.*

Last summer, twelve Quonkians (as the island's residents are called) emigrated to Montgomery, Naomi and Marty's hometown. When the two children first met their new neigh-

bors, they found it difficult to tell the Quonkians apart, because they all looked very much alike. Naomi and Marty were eager to know which of the Quonkians were Trolls, and which ones were Ffaffnirs, so that they might know with whom they could place their trust, and with whom they could not. They asked each Quonkian, in turn, which of the two tribes they belonged to. These were the twelve replies:

Ohm: "Vroom and Won are Trolls."
Pi: "Ohm and Zap are Trolls."
Quid: "Pi and Ung are Trolls."
Rho: "Soy and Zap are Trolls."
Soy: "Quid and Vroom are Trolls."
Typ: "Rho and Won are Trolls."
Ung: "Soy and Xit are Ffaffnirs."
Vroom: "Typ and Yog are Ffaffnirs."
Won: "Ung and Yog are Ffaffnirs."
Xit: "Ohm and Quid are Ffaffnirs."
Yog: "Rho and Typ are Ffaffnirs."
Zap: "Pi and Xit are Ffaffnirs."

From these responses, Naomi and Marty were quickly able to identify which Quonkians belonged to which tribe. Thus, they now knew which ones to believe and which ones they should never trust. Do you?

To simplify matters, when you are solving this puzzle, refer to the Quonkians by the first letters of their respective names, O through Z inclusive, and the tribes by the first letters of their names, T or F. Refer to their statements as either true or false. Notice that statements by T's are always true, and statements by F's are always false. When you have solved the entire puzzle, you can then substitute the names for the corresponding letters.

37 □ Davey is on his way to Meredith's house. He has just come to a brook, which is too wide to jump over, and is too deep to wade across. As you can see in the picture on this page, there is a small island in the middle of the brook, and there are five fallen logs. Two of the logs cross a part of the brook, from the

side where Davey is to the island. Two more cross a part of the brook from the island to the side where Meredith lives. The last log goes directly from Davey's side of the brook to Meredith's.

Just for the fun of it, Davey has decided that he will go to the other side of the brook by walking across every fallen log once and only once. When Davey got to the other side he thought to himself, "That was easier than I thought it would be. I wonder how many different routes there are, from which I could have chosen?" Do you know?

Trace the illustration showing the brook, the island, and the five logs onto a piece of paper. Now, with a pencil, draw as many completely different routes as you can that will take you across each of the five logs once, and only once. You must be careful not to leave out any of the logs. You must also be sure not to count the same route more than once.

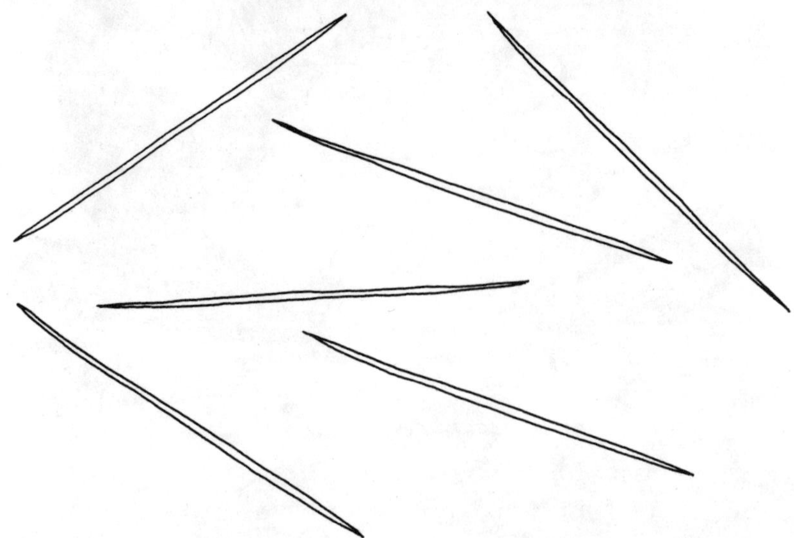

38 □ Arrange 6 toothpicks, of equal length, so that they form 4 equilateral triangles.

39 □ Connie is reading a letter from Sam, who is away on a hunting trip somewhere in the Northern Hemisphere. Sam says in his letter that very early one morning he left his camp and

walked due South for one mile. There he came across some bear tracks, which were heading in an Easterly direction. Sam followed the tracks, and had gone just one mile due East when he overtook and shot the bear. Sam went on to say that after he had shot the bear, he decided to return with it to his camp. Due to the bear's immense weight, he decided to take the shortest route back to his camp. Sam transported the bear back to his camp, in a straight line, from the spot where he had shot it. This third distance that he traveled was also one mile. What color was the bear?

40 □ Draw the irregular shape below on a piece of paper, and cut it out. Now, make a right angle without using a ruler, a protractor, or a pair of compasses, and without copying a right angle from any other figure.

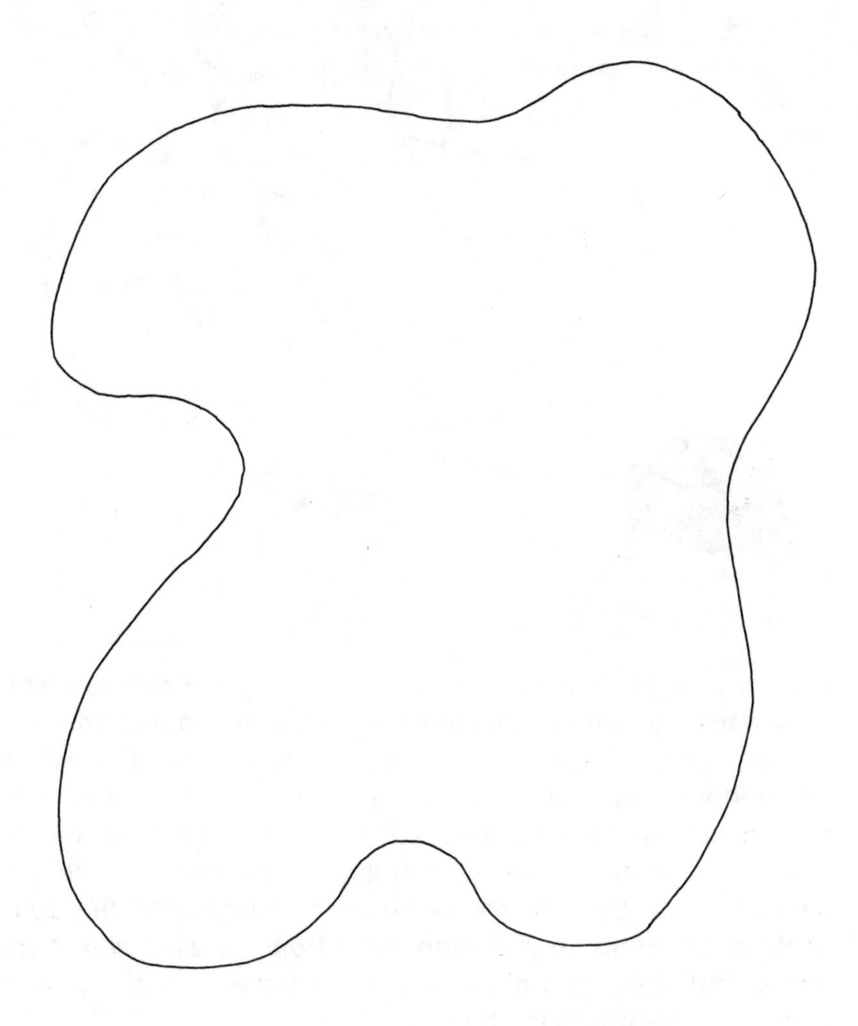

41 □ One day, when Bali and Marty were building a house of cards, Bali happened to glance up toward the ceiling of the room that they were in. There, at a point in the middle of one of the end walls, one foot from the ceiling, she saw an ant. The ant, in turn, was hungrily looking across the room at a crumb, which was peanut-butter-and jellied to the middle of the opposite wall, one foot from the floor.

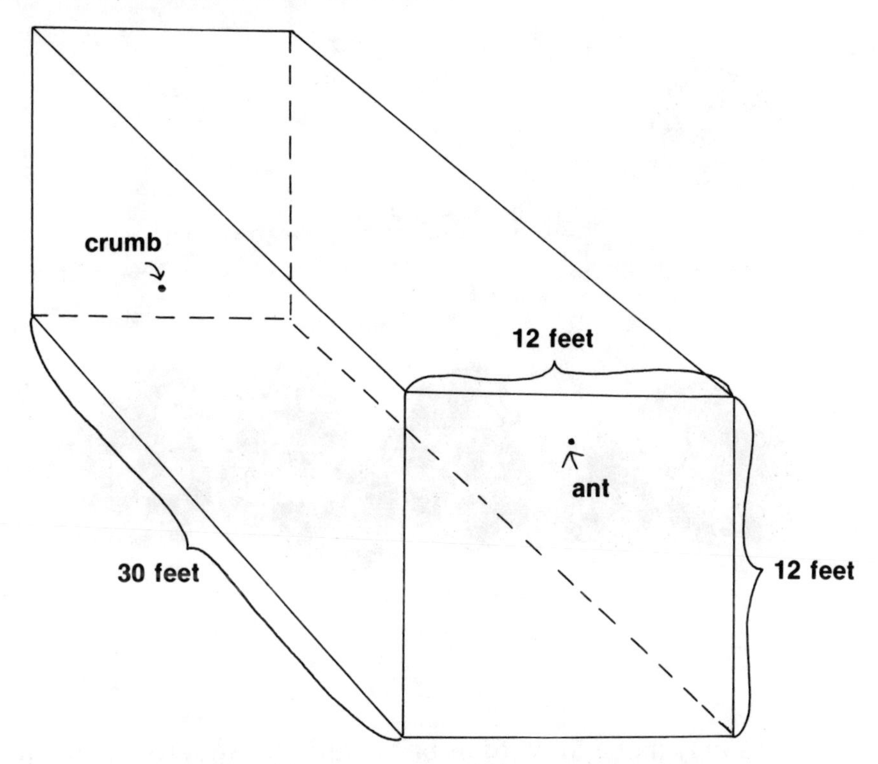

The room that Bali and Marty were in is 30 feet long and 12 feet wide. The distance between the ceiling and the floor is also 12 feet.

What is the shortest distance that the ant will have to crawl in order to reach the crumb?

42 □ Jonathan, Nathaniel, Charlotte, Meredith, Davey, and Amy are standing in a line in six of the seven squares that they have drawn in the dirt. The six hedgehogs would like to reverse the order in which they are standing in the squares, so that the numbers that they are wearing will read 6, 5, 4, 3, 2, 1, and still have the empty square in its original place. Each of the hedge-

hogs may move forward or backward one square, if the square in front or the square behind is empty. Each hedgehog may also leap over one hedgehog in front or one hedgehog behind, into the square that is immediately beyond that hedgehog, as long as that square is empty. What is the fewest number of moves in which the six hedgehogs can reverse their order?

43 □ Jonathan, Meredith, Nathaniel, Charlotte, Davey, and Amy are once again standing in a straight line. This time they are in six of the eight squares that they have drawn in the dirt. The six hedgehogs would like to rearrange their order so that Jonathan, Nathaniel, and Davey are each in any one of the squares 1, 2, and 3, and Meredith, Charlotte, and Amy are each in any one of the squares 4, 5, and 6. The hedgehogs must move two at a time. Each pair of hedgehogs that moves must have been next to each other before they moved. They may only move into empty squares that are next to each other. When a pair of hedgehogs moves, they cannot reverse the order that they were in, from right to left, or left to right, before they moved.

What is the fewest number of moves in which the six hedgehogs can rearrange their order, as described above?

When you are solving this problem, draw eight squares in a row, as in the picture above, on another piece of paper. Use pennies to stand for Jonathan, Nathaniel, and Davey. Use nickels to stand for Meredith, Charlotte, and Amy. Now you can solve the problem by moving the pennies and nickels in pairs, as described above.

44 □ Using commas, colons, periods, or any other punctuation marks, punctuate the following, so that it makes sense:

Ten fingers have I on each hand
Five and twenty on hands and feet

ANSWERS

1 □ There are 3 feet in one yard. Therefore, the number of yards in the trail that Wendy's skis made when she skied down to the bottom of the hill is 360 divided by 3, or 120.

Every minute that Wendy climbed, she moved forward 3 yards, and slipped back 2 yards. Therefore, she actually went up the hill at the rate of 1 yard per minute. At the end of the 117th minute, Wendy had climbed 117 yards up the hill. Thus, at the end of 117 minutes, she was only 3 yards from the top of the hill. She climbed these 3 yards during the next 30 seconds. Therefore, Wendy reached the top of the hill in 117-1/2 minutes, or 1 hour and 57-1/2 minutes.

2 □ The answer to this puzzle is the same no matter how long the Blue Dot Trail is from the bottom of Mt. Minuit to its top. Let us say that the length of the trail up Mt. Minuit is 12 miles. Then the round-trip distance up and down the mountain is 24 miles. Since Davey hiked up the mountain at a rate of 3 miles per hour, it took him 4 hours to reach the top of the mountain. Davey went back down the mountain at a rate of 6 miles per hour, so the descent took him 2 hours. Therefore the total time that it took for Davey to make the round trip was 6 hours. Then the average speed for the round trip is equal to 24 miles divided by 6 hours, or 4 miles per hour. You do *not* obtain the average speed by adding 3 and 6, and then dividing by 2. To get the average speed for the round trip, you must divide the *total* distance by the *total* time.

3 □ The answer to this problem is the same no matter what the distance is between Montgomery and Shaftsbury. Let us say that the distance between the two towns is 120 miles. Then the round trip would be a total distance of 240 miles. If the average speed of the round trip is to be 60 miles per hour, then it should take 4 hours for the bus to make the entire trip. But, if the 120 miles from Montgomery to Shaftsbury are traveled at an average speed of 30 miles per hour, then this trip alone uses up the entire 4 hours, and there is no time left for the bus to make the return trip. Therefore, the bus would have to make the return trip from Shaftsbury to Montgomery in no time at all, which, of course, is impossible. There is no speed for the return trip that can make the average speed for the round trip 60 miles per hour.

4 □ When Amy reached the first sign, 40 minutes after she had left Branford, she had already gone "half as far as it was from there to Guilford." The distance from there to Guilford was double the distance she had already traveled and would take double the time she had already traveled. Therefore, hiking at her own steady pace, from the first sign to Guilford, took Amy another 80 minutes. Therefore it took Amy 120 minutes, or 2 hours, to hike from Branford to Guilford along the Hammonasset Trail.

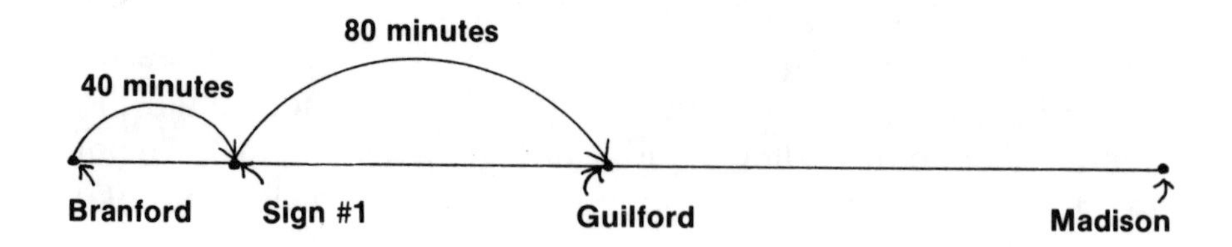

When Amy reached the second sign, she was 11 miles from the first sign. And the distance from the second sign to Madison was "half as far as it was from there to Guilford." Thus, when Amy reached the second sign, she had already passed Guilford and had hiked 2/3 of the distance from Guilford to Madison. Therefore, since the remaining 1/3 of the distance between Guilford and Madison took Amy 1 hour, then the total time that it took for her to hike from Guilford to Madison at her own steady pace was 3 hours.

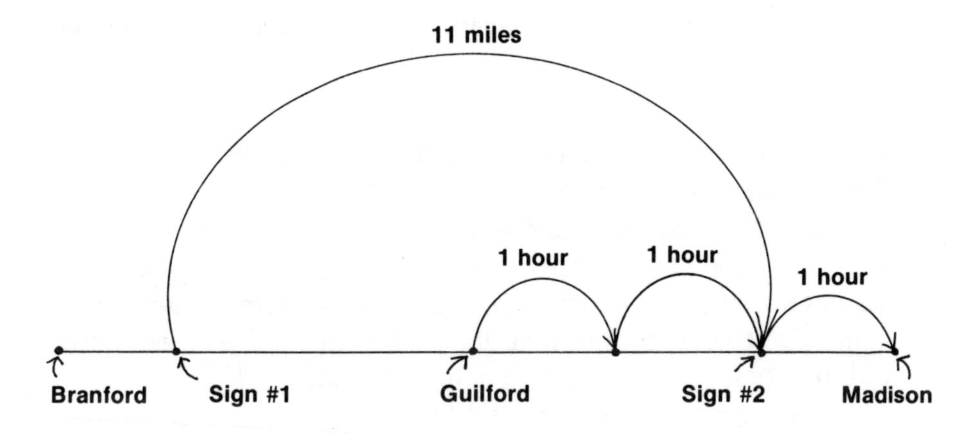

Thus, the total time that it took for Amy to hike from Branford to Madison along the Hammonasset Trail was 2 hours + 3 hours = 5 hours = 300 minutes.

Forty minutes of this time was before she reached the first sign, and 60 minutes of this time was after she passed the second sign. So Amy hiked the 11-mile stretch between the two signs in 300 minutes − 100 minutes = 200 minutes. Thus, hiking at the same steady pace, the other 100-minute part of the trek must have covered 5-1/2 miles of trail. Therefore the total distance that Amy hiked along the Hammonasset Trail, from Branford to Madison, was 11 miles + 5-1/2 miles, or 16-1/2 miles.

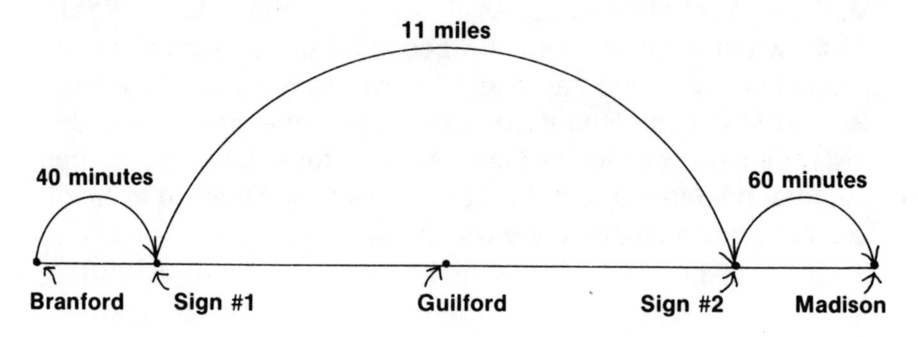

5☐ The riverboat that Wendy was on arrived back at the Canal Street dock in New Orleans at 4:32 P.M.

The distance that the riverboat traveled along the Mississippi River, when the bayou tour began, was 10 miles. The distance that the boat traveled along the Algiers Route was 7 miles. Therefore, by the time the riverboat had reached Hero Cutoff, it had already gone 10 miles + 7 miles, or 17 miles.

The riverboat then traveled the 14 miles along the Bayou Barataria, turned around, and immediately returned the 14-mile distance to Hero Cutoff. Thus, the total distance that the riverboat traveled along the Bayou Barataria was 14 miles + 14 miles, or 28 miles. Therefore, when the riverboat entered the Harvey Canal, it had already gone 17 miles + 28 miles = 45 miles.

The distance that the riverboat traveled along the Harvey Canal was 3 miles. It then turned back onto the Mississippi River and traveled the final 4 miles to the Canal Street dock in New Orleans. Therefore, the total distance of the bayou tour was 45 miles + 3 miles + 4 miles = 52 miles.

The riverboat traveled at an average speed of 10 miles per hour. Since Time = Distance divided by Rate, the boat's traveling time was $= \frac{52 \text{ miles}}{10 \text{ miles per hour}} =$ 5 hours and 12 minutes.

The riverboat also spent 10 minutes in the Algiers Lock, and 10 minutes in the Harvey Lock. Thus, the total time that it spent in the two locks was 10 minutes + 10 minutes, or 20 minutes. Now add the 20 minutes that the riverboat spent in the two locks to the 5 hours and 12 minutes that it took for it to travel the 52 miles of the tour. 5 hours and 12 minutes + 20 minutes = 5 hours and 32 minutes. Thus, the riverboat returned to the Canal Street dock 5 hours and 32 minutes after it had departed.

The bayou tour began at 11 A.M. 11 A. M. + 5 hours and 32 minutes = 4:32 P.M.

6 □ Since light travels 186,000 miles per second, it took almost no time at all to travel the short distance between Bali and Marty. So we can assume that Marty *saw each swing* of the hammer at the time that Bali made it. However, he didn't hear the bang of the swing until after Bali made the swing, because he didn't hear the sound until after it traveled the distance between him and Bali. To figure out what this distance was, we observe:

1. Bali hit a nail 30 times every minute. There are 60 seconds in a minute. So Bali hit a nail every two seconds. Thus there are two seconds between each swing of the hammer and the swing just before it.
2. The last bang Marty heard was the bang of the last swing. When he saw the last swing of the hammer, it was accompanied by a bang. That bang was the bang before the last, so it was the sound of the swing before

the last. Each swing was separated from the one before it by two seconds. So it was two seconds between the time Marty saw the last swing and the time when he heard its bang. This means that it took the sound of the bang two seconds to travel from Bali to him. At the rate of 1,100 feet per second, the sound traveled 2,200 feet in two seconds. So the distance from Bali to Marty was 2,200 feet.

7 □ Because Amy ran the first half-mile and the second half-mile each in the same amount of time, the times she took for the first half-mile and for the second half-mile were each one-half of the time she took for the whole mile. Because she ran the third quarter and the last quarter each in the same amount of time, this time was half of the time taken for the last half-mile, so it was 1/4 of the time for the whole mile. Then the time taken for the first three-quarters of a mile was $(1/2 + 1/4)$ or 3/4 of the time for the mile. Then 6-3/4 minutes is 3/4 of the time taken for the mile. $6\text{-}3/4 = 27/4$. If 3/4 of the time was 27/4 minutes, then 1/4 of the time was $27/4$ minutes $\div 3 = 9/4$ minutes. Then the whole time was $9/4$ minutes $\times 4 = 9$ minutes.

8 □ Charlotte will have picked the larger crop of strawberries. She picked six dozen dozen strawberries, which is the same thing as $6 \times 12 \times 12$, or 874 strawberries. Amy picked a half dozen dozen strawberries, which is the same thing as 6×12, or 72 strawberries.

9 □ When Davey made his promise to Amy, he had intended to give her 2 times as many mushrooms as he would keep for himself.

When Davey made his promise to Meredith, he had intended to give her 1/2 as many mushrooms as he would keep for himself.

Therefore, Amy was supposed to receive 2 shares, to Davey's 1 share, to Meredith's 1/2 share. The ratios will be the same if we double each of these numbers.

Thus, Amy should receive 4 shares of the mushrooms, Davey should keep 2 shares of the mushrooms, and Meredith should receive 1 share of the mushrooms.

This, then, gives us a total of 7 shares of mushrooms to be divided among the three friends. Therefore, Amy received 4 of the 7 shares of the mushrooms, or 4/7 of Davey's day's pick. Davey, himself, kept 2/7 of his own day's pick, and Meredith received the remaining 1/7 of the mushrooms.

10 □ First you find the whole number that is the sum of the four fractions. Every fraction whose numerator is one less than its denominator is *less than one.* Therefore, the sum of four such fractions will be less than four.

The smallest fraction whose numerator is one less than its denominator is 1/2. Therefore, since the four fractions are different, their sum must be greater than 4 times 1/2, or 2. The only whole number that is greater than 2, yet is less than 4, is 3. Therefore, the sum of the four fractions is 3.

Now you are ready to find four different fractions, of the kind required, whose sum is three. Start with the fraction 1/2, and subtract it from 3. 3 – 1/2 = 5/2. The next fraction to try is 2/3. Subtract it from 5/2. 5/2 – 2/3 = 11/6. If we use 1/2 and 2/3 as two of the four fractions, the other two must add up to 11/6. 3/4 can't be one of them, because 11/6 – 3/4 = 13/12, which is greater than 1, which is more than the fourth fraction can be. 5/6 can't be one of them, because 11/6 – 5/6 = 6/6 = 1.

So the next larger fraction worth trying is 7/8. 11/6 − 7/8 = 23/24, which has the required form. So one answer to the question is, 1/2 + 2/3 + 7/8 + 23/24 = 3. There are several other possible answers, too.

11 □ It takes 49 minutes to cut off 49 inches. The remaining inch, or fiftieth inch, does not need to be cut. Therefore, it takes 49 minutes to cut the entire 50-inch strip into one-inch pieces.

12 □ It is possible to divide the 3 apples among the 2 fathers and 2 sons, so that each receives exactly 1 apple, if the persons receiving the apples are: a man, his son, and his grandson. Thus, there are 2 fathers in the group and there are 2 sons in the group, but there are only 3 persons.

13 □ Five hedgehogs will eat 100 mushrooms in 100 minutes. The five hedgehogs as a team eat mushrooms at the rate of 1 mushroom per minute. At this rate, they can eat 100 mushrooms in 100 minutes.

14 □ There are three hedgehogs, and there are 21 jugs. So for each hedgehog to receive the same number of jugs, each should get 7 jugs.

In order for each of the three hedgehogs to receive an equal amount of molasses, let us see how many jugs-full of molasses there are in all. There are 7 full jugs. There are also 3 half-full jugs and 4 half-empty jugs. A jug that is half-full contains the same amount of molasses as a jug that is half-empty. Thus, there are 7 jugs that are half-filled with molasses. Seven half-filled jugs of molasses is the same thing as 3-1/2 jugs-full of molasses. Therefore, Jonathan has a total of 7 full jugs of molasses + 3-1/2 jugs-full of molasses = 10-1/2 jugs-full of molasses. Ten and a half jugs-full of molasses divided among 3 hedgehogs will give each one 3-1/2 jugs-full of molasses.

Jonathan divided his 21 jugs among Nathaniel, Meredith, and himself so that each one received his or her 3-1/2 jugs-full of molasses, and his or her 7 jugs, in the following manner:

	FULL	HALF-FULL	HALF-EMPTY	EMPTY
Nathaniel	3	0	1	3
Meredith	2	3	0	2
Jonathan	2	0	3	2

15 □ The flask and the jug each contained one pint of fluid before any was transferred from one to the other. Each also contained one pint of fluid after the transfers. So, whatever the amount of juice that went from the flask to the jug and then stayed in the jug, an equal amount of cider must have gone from the jug to the flask to restore the original volume of one pint. So neither the flask nor the jug received more than it gave.

16 □ Three-quarters of a pound is equal in weight to 1/4 of Paxi's weight when she is wet. Thus, when Paxi is wet she weighs 4 times 3/4 of a pound. $4 \times 3/4 = 12/4 = 3$ pounds. Therefore, when Paxi and Blixi are dry, they each weigh 3 pounds minus 3/4 of a pound = 2-1/4 pounds.

17 □ When you take any three-digit number, abc, and convert it to abcabc, it is the same thing as multiplying the original three-digit number by 1001, and $1001 = 7 \times 11 \times 13$.

18 □ $\frac{3}{.3} \times \frac{3}{.3} =$

$$\frac{3}{\frac{3}{10}} \times \frac{3}{\frac{3}{10}} =$$

$$\left(3 \times \frac{10}{3}\right) \times \left(3 \times \frac{10}{3}\right) =$$

$$\frac{30}{3} \times \frac{30}{3} =$$

$$10 \times 10 = 100$$

19 □ $9 + 99/9 =$

$9 + 11 = 20$

20 □ There are many even numbers that can be written using only odd digits. Some of them are: 3-3/3, 5-5/5, 7-7/7, etc.

21 □ A hundred can be written without using zeros in a variety of ways: 99 + 9/9, C, hundred.

22 □ It takes 4 × 4, or 16 quarter-inch squares to make an inch square.

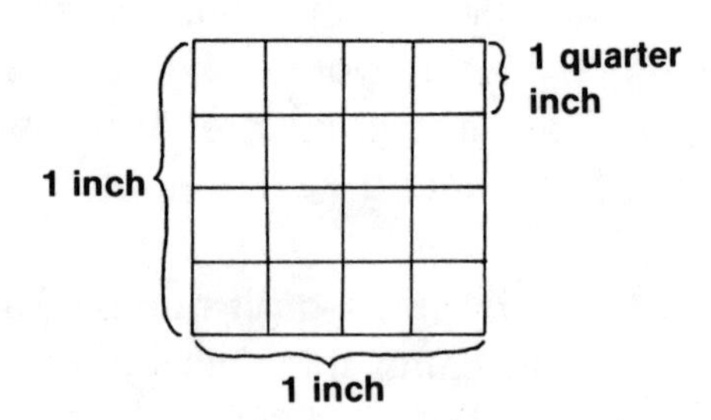

23 □ First, you must figure out what sum each individual row should add up to. The sum of all of the numbers, from 1 to 9, inclusive, is 45. Since there are three rows in the square, then each of these rows must add up to 45 divided by 3, or 15.

Here are two arrangements for a 3 by 3 magic square. You can obtain other arrangements by writing the numbers in each of the rows backwards, or by reversing the order of the numbers in each of the columns.

8	1	6
3	5	7
4	9	2

6	7	2
1	5	9
8	3	4

24 □

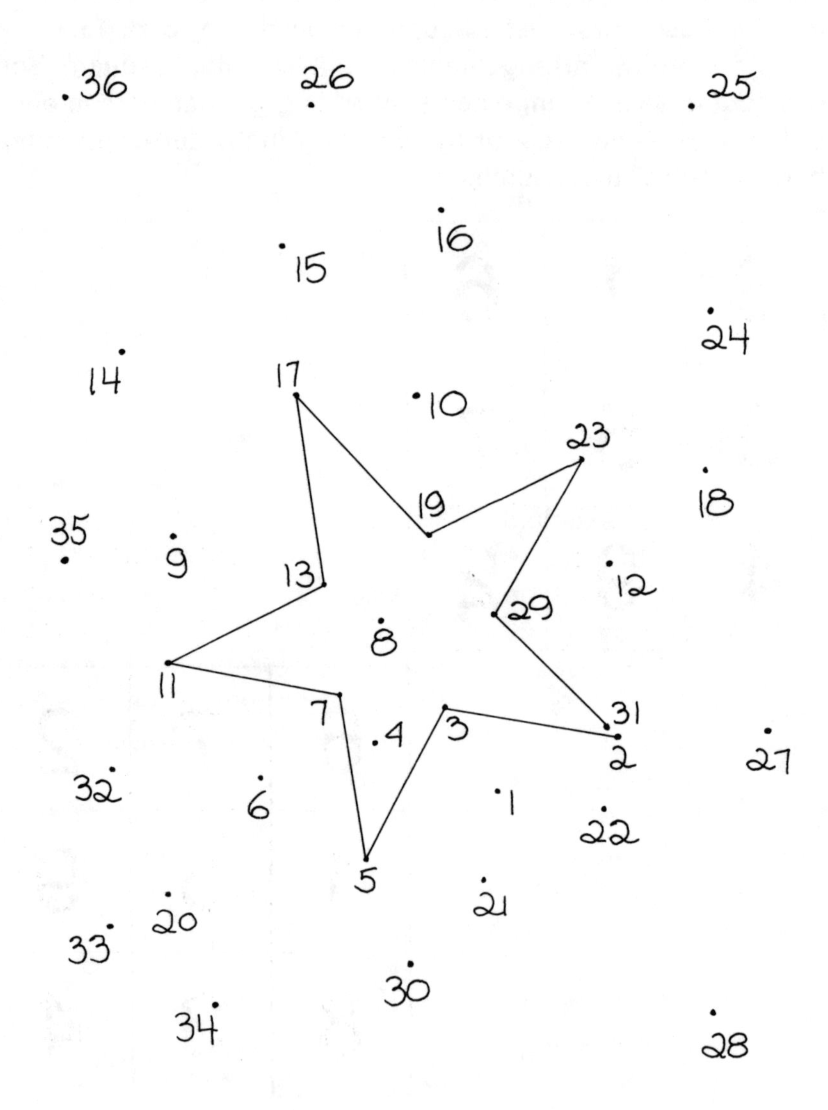

36
26
25
16
15
24
14
17
10
23
18
19
35
9
13
12
29
8
11
7
31
3
4
2
27
32
6
1
22
5
21
20
33
30
34
28

25 □ First consider Connie and Sam's ages after their next birthdays.

Connie's age then cannot be more than 11, since the square of 12 is *already* more than 135. Thus, the square of Sam's age is at least 207 – 11 = 196, and 196 is the square of 14. Therefore, try the ages 11 and 14 first:

$$11 + 14^2 = 11 + 196 = 207.$$
$$11^2 + 14 = 121 + 14 = 135.$$

These numbers fit the conditions of the problem. Therefore, Connie's age *after* her next birthday is 11, and Sam's age *after* his next birthday is 14. Thus, Connie's present age is 10, and Sam's present age is 13.

26 □ When a number is written as the product of two factors, the two factors can be equal only if the number is a square number. For example, 12, which is not a square number, may be written in these ways as the product of two factors: 1×12, 2×6, 3×4. In each pair the factors, or divisors, are different, so there must be an even number of them. But 16, which is a square number, may be written as the product of two factors in these ways: 1×16, 2×8, 4×4. In the last pair, the two factors are equal. When we count the factors, or divisors, we count the 4 only once. So the number of divisors of 16 is odd.

In fact, the *only* numbers that have an odd number of divisors are square numbers. The only square number between 50 and 79 is 64. Therefore, Naomi and Marty's grandfather will be 64 on his birthday this year.

27 □ Bali's refreshment stand operated at a loss, no matter how large her total sales were.

Suppose, for example, her iced tea sales and her lemonade sales were both equal to 99¢. Her lemonade sales produced a profit of 10%. Then the 99¢ she received for the lemonade was 90¢ cost plus 9¢ profit (1¢ of profit for each 10¢ of cost). Her iced tea sales produced a loss of 10%. Then the 99¢ she received for the tea was 110¢ cost minus 11¢ loss (1¢ of loss for each 10¢ of cost). The combined profit and loss of the two transactions is 9¢ profit + 11¢ loss = a loss of 2¢.

28 □ If you said 20%, then you are wrong! Let us say that Sam's new jeans cost $10 before they were put on sale. Then, one single discount of 20% would be equal to 20% times $10 = .20 × $10 = $2. Therefore, Sam would have bought the pants for $8 if there had been a single discount of 20%.

Now let us take two successive 10% discounts on the same pair of pants. We said that the original price of the pants was $10. Therefore, 10% of $10 = 10% times $10 = .10 × $10 = $1 was the amount of the first 10% reduction. Then $10 – $1 = $9 equals the price of the pants after the first 10% reduction. Now make the second 10% reduction. 10% of $9 = 10% times $9 = .10 × $9 = 90¢ is the amount of the second discount, after the second 10% reduction. Therefore, Sam bought his new jeans for $9 – 90¢, or $8.10. This price is 10¢ more than he would have paid had there been one single 20% discount. Sam's total price reduction was $1 + 90¢ = $1.90, which is only 19% of $10. Therefore, two successive 10% discounts are the same as one single 19% discount.

29 □ After the first 3 years, there was an increase of half of $2,000, or $1,000. Thus, on Connie's third birthday, her account had $2,000 + $1,000, or $3,000 in it. After 6 years, there was an increase of half of $3,000, and the new total was $4,500. After 9 years, the total was $6,750; after 12 years, $10,125; after 15

years, $15,187.50; and after 18 years, there was $22,781.25 in Connie's college fund.

30 □ The least amount of money that Connie could have spent to have the seven small chains linked together to form one long chain is $3.75.

First the silversmith cut open each of the five links in one of the seven small chains. Since he charges 25¢ for each link that he cuts open, the cost of opening the five links is 5 times 25¢ = $1.25.

Then the silversmith arranged the remaining six small chains in a line. There were five places in the line where each one of the small chains lay next to another. In each of these places the silversmith used one of the open links to join together the neighboring chains. Since he charges 50¢ for each link that he solders, the cost of closing each of the five links is 5 × 50¢ = $2.50.

Therefore, the total cost of opening and closing each of the five links is $1.25 + $2.50 = $3.75.

31 □ The largest sum of money that Bali could have had in her pockets was $1.19: a half-dollar, one quarter, four dimes, and four pennies.

32 □ There are 100 pennies in $1. Therefore, Charlotte has 100 more pennies than Amy. Thus, Charlotte has 105 pennies, and Amy has 5 pennies.

33 □ The cost of one ice cream cone and one banana split is the same as the cost of 4 ice cream cones. Therefore, the total amount of their purchase must be a number that can be divided evenly by 4.

We already know that the children had one dollar between them, and that they had some money left over after they had made their purchases. Therefore, the total cost of the ice cream cone and the banana split must have been less than one dollar. Thus, there are two digits in the total price. The only pairs of digits whose sum is 14 are 9 and 5, 8 and 6, and 7 and 7. The two-digit numbers they can form are 95, 59, 86, 68, and 77. The only one of these that is divisible by four is 68.

Therefore, the total cost of their purchases was 68¢. 68¢ ÷ 4 = 17¢ is the cost of one ice cream cone. Thus, the banana split cost 68¢ – 17¢ = 51¢.

34 □ The train that Wendy was on left New York at 2 P.M. At the same time that Wendy's train pulled out of the station in New York, a train that had left New Haven at noon pulled into the station. This was the first train that Wendy saw coming from the opposite direction. Just as Wendy's train pulled into the station in New Haven, at 4 P.M., the 4 P.M. train to New York pulled out. This was the last train that Wendy saw going in the opposite direction. Therefore, Wendy saw trains that left New Haven at noon, 1 P.M., 2 P.M., 3 P.M., and 4 P.M. Thus, she saw a total of 5 trains.

35 □ The face of the clock is divided into 60 equal spaces to represent the 60 minutes in an hour. The hour hand travels clockwise at a speed of 5 spaces per hour. The minute hand travels clockwise at a speed of 60 spaces per hour, so it moves away from the hour hand at a speed of 60 spaces – 5 spaces = 55 spaces an hour. The minute hand and the hour hand will be together again when the gap opening up between them at this rate is 60 spaces long. The time this takes in hours is 60/55 = 1-1/11. Since 1/11 of 60 minutes is 5-5/11 minutes, the second

showing begins at 5-5/11 minutes past 1 P.M.

36 □ To solve this puzzle, look for statements that lead to conclusions that contradict each other or are known to be false. Such statements are then known to be false (f), and must have been made by an F.

1. If P is a T, P's statement is true, and then Z is a T. If Z is a T, Z's statement is true, and P is an F. Then "P is a T" leads to the contradictory statement "P is an F." Therefore, "P is a T" is false, and then *P must be an F.*
2. If Q is a T, Q's statement is true, and P is a T. But we know now that P is not a T. Then Q's statement is false, and *Q is an F.*
3. If S is a T, then S's statement is true, and Q is a T. But we now know that Q is not a T. Then S's statement is false, and *S is an F.*
4. If R is a T, then R's statement is true, and S is a T. But we now know that S is not a T. Then R's statement is false, and *R is an F.*
5. If T is a T, then T's statement is true, and R is a T. But we now know that R is not a T. Then T's statement is false, and *T is an F.*
6. Y says that R and T are F's, which is true. Then *Y is a T.*
7. V says that T and Y are F's, which is not true. Then *V is an F.*
8. O says that V and W are T's, which is not true. Then *O is an F.*
9. X says that O and Q are F's, which is true. Then *X is a T.*

10. Z says that P and X are F's, which is not true. Then *Z is an F.*
11. U says that S and X are F's, which is not true. Then *U is an F.*
12. W says that U and Y are F's, which is not true. Then *W is an F.*

37 □ A simple way to identify the different routes that Davey can take is to use the letters v, w, x, y, and z for the five logs, as shown in the drawing. D stands for Davey's side of the brook; M stands for Meredith's side; and I stands for the island. Davey can use the island in two ways. Each time he reaches the island from one side of the brook, he can either 1) cross over immediately to the other side of the brook; or 2) return immediately to the same side. In case 1, a crossing from D to M or from M to D via the island uses two logs. A crossing that does not use the island uses only the log z. To use each of the five logs once and only once, Davey must cross the brook three times, twice via the island, and once via log z. He has three ways of starting, because he can use first either log v or log w or log z. The different sequences of logs he could take are these:

If he starts with v, and next takes x: vxywz, vxzwy.
If he starts with v, and next takes y: vyxwz, vyzwx.
If he starts with w, and next takes x: wxyvz, wxzvy.
If he starts with w, and next takes y: wyxvz, wyzvx.
If he starts with z, and next takes x: zxwvy, zxvwy.
If he starts with z, and next takes y: zyvwx, zywvx.

In case 2, the possible paths are:

vwzxy, wvzxy, vwzyx, and wvzyx.

The number of different paths is 16.

38 □ You are not required to put all six toothpicks flat on the table. You can use them to form a triangular pyramid, as shown in the drawing

pyramid, as seen from above

39 □ Draw a diagram of the distances that Sam said that he went. Your diagram will be of a triangle with 3 equal sides, or an equilateral triangle. The only place in the Northern Hemisphere where it is possible to go due South for one mile, then due East for one mile, and finally return to the starting point by going, again, one mile, or in an equilateral triangle, is at the North Pole. And, the only kind of bear that is found at the North Pole is a polar bear. Therefore, the bear that Sam shot was white.

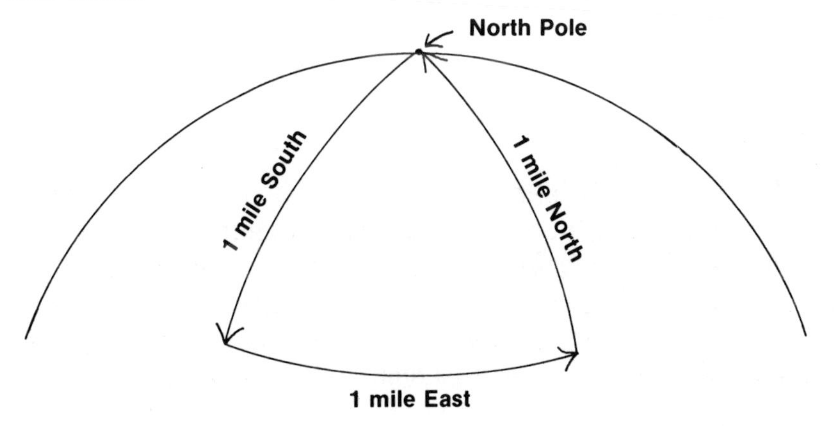

40 □ Fold the paper once and press the crease flat, as shown in diagram #1. The crease will be a straight line. Then fold the paper again so that one part of the crease lies on the other part, as shown in diagram #2.

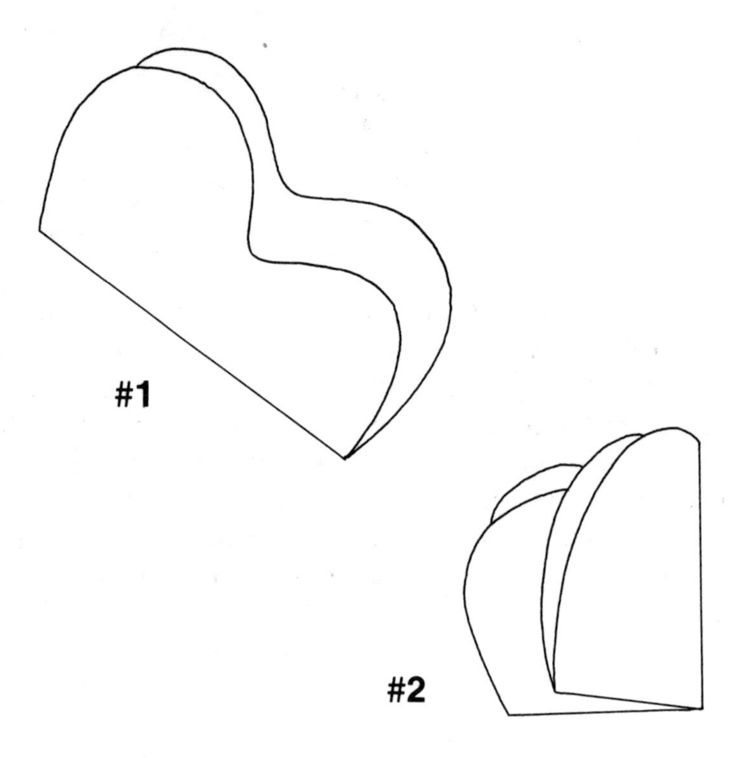

41 □ The room that Bali and Marty are in is a rectangular box. You can make such a box out of a piece of paper. This box may be cut open in a variety of ways so that the paper can be laid flat upon a table. For each different way that you cut the box open, indicate the ant and the crumb by dots, in their respective positions on the two end walls. Each time that you cut

open the box in a different way, draw a line between the two dots, without going off the paper. In each case, this is a route that the ant may take. One such route is shorter than any other, and is represented by the dotted line in the diagram below.

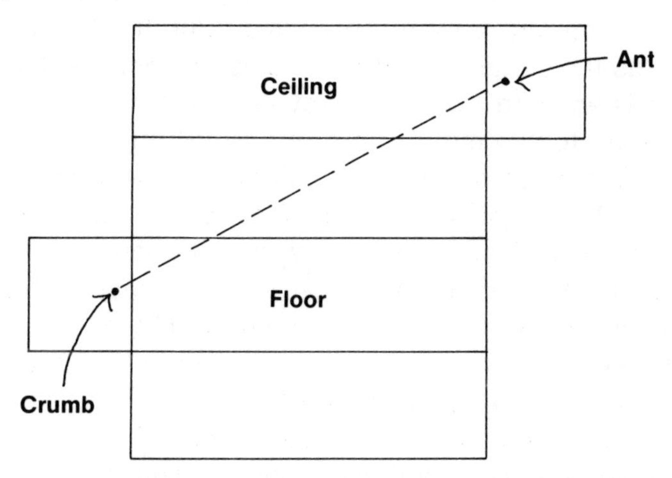

This distance that the ant traveled is the hypotenuse of a right triangle, as you can see in the next diagram.

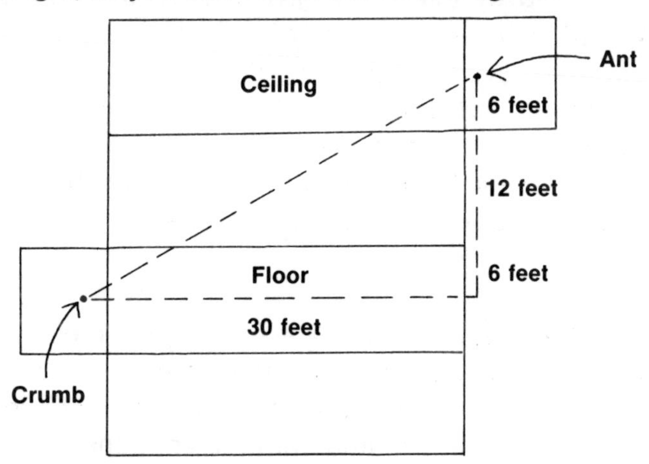

The length of one of the legs of the right triangle is equal to one-half of the width of the end wall that the ant is on, plus the height of the room, plus one-half of the width of the floor. This distance is equal to 6 feet + 12 feet + 6 feet, or 24 feet. The length of the other leg of the right triangle is equal to the distance that the ant is from the ceiling, plus the length of the room, plus the distance that the crumb is from the floor. This distance is equal to 1 foot + 30 feet + 1 foot, or 32 feet.

In order to find the length of the hypotenuse of a right triangle, you must do the following: First, you must *square* each of the legs. When you square a number you are multiplying that number by itself. We already know that the length of one leg of this particular right triangle is 24 feet, and that the length of the other is 32 feet. Thus, 24 feet squared = $(24 \text{ feet})^2$ = 24 feet × 24 feet = 556 square feet. And, 32 feet squared = $(32 \text{ feet})^2$ = 32 feet × 32 feet = 1024 square feet. Now you add the squares of these two sides together. 556 square feet + 1024 square feet = 1600 square feet. Next, you must find the *square root* of 1600 square feet. The square root of 1600 is that number, which when multiplied by itself, will give you 1600. Thus, since 40 times 40 = 1600, the square root of 1600 square feet = 40 feet = the hypotenuse of this right triangle. Therefore, the shortest distance that the ant would have to crawl around the room, to reach the crumb, is 40 feet. To verify that this is the shortest path, calculate the length of the line joining the ant to the crumb when the box is opened flat in other ways.

42 □ Write the numbers from one to six on a piece of paper. Cut each of the numbers out and tape them, one at a time, onto each of six pennies. Now draw seven squares in a row, as in the picture on page 63, on another piece of paper.

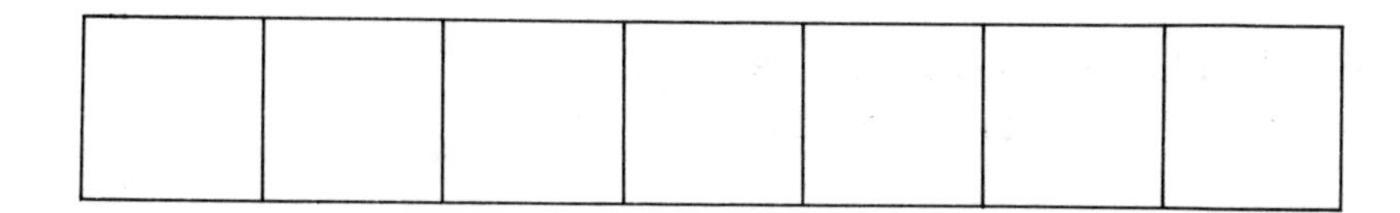

The best way to solve this puzzle is to try first an easier version of it. Play the game with 2 pennies and 3 squares. Leave the first square blank, and put pennies 1 and 2 on the others in the order 1, 2. Use the fewest possible moves to produce the order 2, 1, with the blank square finally ending up on the left. You should be able to do it in 3 moves. Then play the game with 4 pennies and 5 squares. You should be able to reverse the order of the pennies in 10 moves. Now play the game with all six pennies and seven squares. You should be able to reverse the order of the pennies in 21 moves. One way of doing it is to move the pennies in this order: 2, 4, 6, 5, 3, 1,
2, 4, 6, 5, 3, 1,
2, 4, 6, 5, 3, 1,
2, 4, 6.

43 □ The least number of moves in which the hedgehogs can rearrange their order is 4. Charlotte and Davey move to squares 7 and 8 respectively. Now squares 4 and 5 are empty. Amy and Charlotte now move to squares 4 and 5 respectively. This move now leaves squares 6 and 7 empty. Meredith and Nathaniel now move into squares 6 and 7 respectively, leaving squares 2 and 3 empty. The final pair to move will be Nathaniel and Davey. They will move into the empty squares 2 and 3 respectively. Now Jonathan, Nathaniel, and Davey are in squares 1, 2,

and 3, and Amy, Charlotte, and Meredith are in squares 4, 5, and 6.

44 □ Ten fingers have I; on each hand
Five; and twenty on hands and feet.